Revise and Test

GW00707737

Statistics 1

Paul Whitehead PhD MSc

Series Adviser: Geoffrey Whitehead BSc(Econ)

Pitman

PITMAN PUBLISHING LIMITED
128 Long Acre, London WC2E 9AN

PITMAN PUBLISHING INC
1020 Plain Street, Marshfield, Massachusetts 02050

Associated Companies
Pitman Publishing Pty Ltd, Melbourne
Pitman Publishing New Zealand Ltd, Wellington
Copp Clark Pitman, Toronto

First published in Great Britain 1985

British Library Cataloguing in Publication Data
Whitehead, Paul
  Statistics.—(Revise and test)
  1. Mathematical statistics—Problems, exercises, etc.
  I. Title   II. Series
  519.5'076   QA276.2

  ISBN 0–273–02331–4

Typeset and printed in Great Britain at The Bath Press, Avon

Contents

Using this Revise and Test booklet

1   The 'Revise and Test' series is in question and answer form. It will help you revise everything you need to known about your particular syllabus. The questions are detailed and rigorous, and cannot be answered always with one word. It follows that the first time you go over a topic you may be learning the material rather than testing yourself. It is not just a self-testing book, but a self-teaching book too!

2   The first time you study a topic you may need to go over it two or three times. Then put a tick against the topic number in the check list at the back of the book.

3   Subsequently you should revise the topic at intervals, especially just before a monthly test or an examination. Each time you revise it put in a further tick.

4   If you find a topic particularly difficult, put a ring round the number. This will remind you to do it again soon. Practice makes perfect.

5   Finally remember that learning facts is relatively easy. Applying them in written work is more difficult. Each topic has one piece of written work and you should find others from textbooks and past examination papers. Remember the saying 'Writing maketh an exact man'. Don't worry about who is going to mark your written work. You can appraise it for yourself! Keep writing!

By the same author
Statistics for Business, Pitman, 1984

1 What is statistics?

1 How would you define statistics?

As numerical facts, systematically collected and presented.

2 What is the origin of the word?

The word 'statistics' is derived from the Latin for 'state'. Governments have always been interested in the numbers of their citizens and in trade figures, etc.

3 Who else is interested in statistics?

Managers in many fields of endeavour such as business, economics, social science, science and engineering also require numerical facts or data. Sales figures, interest rates, instrument measurements or metal fatigue figures provide facts on which management decisions can be based.

4 What is meant by statistical method?

A range of mathematical techniques for analysing problems in the real world. The mathematical techniques are averages, trends, probability distributions, etc., and these can be used to test ideas and hypotheses.

5 Give an example of this use of statistics

We might wish to hypothesise that adding nitrogen fertiliser to our crops would improve the yield up to a certain application rate, but beyond this rate the excess nitrogen will be merely flushed out by rainfall into the nearest river. We could use statistical method to determine the optimal application rate.

6 What do we mean by descriptive statistics?

These are raw or untreated data presented in such a manner as to reduce confusion and highlight important features. The data may be sorted and tabulated. For example, a typical set of data concerning college students could be arranged as shown in Question 7.

7 Explain the data set out below

Age	Type of student	Number
16–18	School leavers	460
19–21	Others under 21	956
21–24	Postgraduates	67
24+	Other mature students	53
	Total =	1536

(The reader is left to answer this question.)

8 What are analytical statistics?

Statistics which provide methods of analysis which enable decisions to be made. They are sometimes called *inductive statistics* since they lead us to infer generalisations or predictions from the data we have collected. The decision may be positive, such as a decision to expand production, or a decision to take no action at all.

9 Is statistics a science?

Statistics requires a scientific approach to ensure that we deal objectively with our problems and do not distort statistical methods to meet our own expectations. It is a search for truth, like all science.

10 Is statistics also an art?

Whilst keeping our scientific integrity we must also exercise our ingenuity in presenting statistics. A presentation to a board meeting or a panel of decision makers requires artistic flair if the data is to be presented in an attractive and simple form, and we must always consider our audience in presenting facts, information, recommendations and conclusions.

Written Exercise: *Distinguish between descriptive statistics and analytical statistics. In your answer refer particularly to an investigation of road traffic statistics on a particular single carriageway road which is being considered for up-grading to dual carriageway.*

Go over the topic again until you are sure of all the answers. Then tick it off on the check list at the back of the book.

2 Conducting a statistical inquiry

1 What are the stages of a statistical inquiry?	(a) A clear statement of the problem; to give us terms of reference for the inquiry. (b) A literature or in-house records survey. (c) A 'census or sample' decision. (d) Preparation of a questionnaire. (e) Appointment and training of interviewers. (f) Collection of the data. (g) Checking, editing and classifying the data. (h) Analysis of the data. (i) Presentation and report writing.
2 How are the terms of reference obtained?	An inquiry cannot be launched in general terms. Those instituting the inquiry must identify the cause for concern and state explicitly what the problem is. The area giving trouble may be a particular department, or a particular process or product. Stating the problem carefully gives those conducting the inquiry terms of reference from which they can start to collect relevant data for analysis.
3 How do we start to tackle the problem?	By conducting a literature survey or an in-house-records survey.
4 What are these?	In a 'literature survey' we read all the relevant published material. In an in-house-records survey we discover whether relevant data is already

available from records accumulated in the past. If sufficient data is not available we must commence the collection of data.

5 How do we proceed to do this?

The extent of the inquiry is first decided. Should we obtain data on the whole 'population' as in a national *census*, or from a *sample* of the population?

6 What are the usual methods used to obtain census or sample data?

We may send out a questionnaire or interview a representative sample of people to obtain their views on a particular subject. Special precautions are necessary to ensure that a random sample is interviewed and not a biased sample which will give a biased result.

7 What are the salient points about a questionnaire?

It is the simplest way to conduct an inquiry. The points to remember are:
(a) Questions should inquire about all the data required, in logical sequence.
(b) They should preferably only require a tick, or the encirclement of a number to show the selected answer. (c) There should not be too many questions, as the responder may become bored.
(d) There should be an introductory paragraph to explain the uses and aims of the inquiry. (e) If answering is compulsory the penalty for failure to complete the questionnaire should be stated. (f) Clear instructions about the return of the questionnaire should be given. (g) In the design, the ease of collation of the responses should be borne in mind. (h) The questionnaire is then sent out either to the whole population (the entire group interested in this set of data) or to a representative sample.

8 How does an inquiry by interview differ from the use of a questionnaire?

We can rarely interview the whole population. We have to select a sample (see Topic 5 in this book) and provide the interviewer with a *schedule* of questions which is very similar to a questionnaire.

9 What are the rules about interviews?

(a) The schedule of questions must be comprehensive, and logical in sequence, but not too long. (b) Care must be taken to avoid biased questions and hence biased answers. (c) It must be easy for the interviewer to record the responses. (d) Interviewers must be adequately briefed and instructed.

10 How else can data be collected?

By observation. For example, in the case of a traffic census, enumerators record traffic passing a specific point, using the five-barred-gate principle illustrated in Fig. 2.1. Other data can be collected electronically, by telemetering devices in industrial plant, for example, if the inquiry is about quality control.

Traffic census : Kano Road

Date: 27 July 19 –
Time: 9.00am – 10.00am

Bicycles	Mopeds/motorcycles	Private cars and vans	Lorries and buses
₩₩ ₩₩ ₩₩ ₩₩ ₩₩ ₩₩ ‖ ㉜	₩₩ ₩₩ ₩₩ ₩₩ ₩₩ ₩₩ ₩₩ ₩₩ ‖‖‖ ㊹	₩₩ ₩₩ ₩₩ ₩₩ ₩₩ ₩₩ ₩₩ ₩₩ ₩₩ ₩₩ ₩₩ ₩₩ ‖ ㉚	₩₩ ₩₩ ₩₩ ‖‖‖ ⑱

Fig. 2.1 The five-barred-gate system of enumeration

11 How is this raw data analysed and presented?

The result of data collection is a mass of raw data which first needs editing and classifying. The stages are: (a) Check for obvious errors, joke answers, etc. (b) Collate the material, classifying items to reduce the volume of data to be handled. For example, in Fig. 2.1 the total number of vehicles in each section is obtained, and shown in a circle. These

sub-totals are then carried to a master sheet which will provide the total for each day under each heading.
(c) Derive other statistics from the condensed data, averages, percentages, standard deviations, coefficients of skewness, correlation coefficients, etc., so that a clear picture is obtained and correct conclusions can be drawn. (d) Equally important is a clear, simple and convincing presentation of the results, so that decisions can be made with the correct interpretation of the data. Table, charts and figures will help show survey results, suggest causes or reasons for the original problem and make firm recommendations to resolve the difficulty.

Written Exercise: *What are the problems of conducting a statistical inquiry which requires ratepayers to be interviewed about their opinions of local government services.*

Go over the topic again until you are sure of all the answers. Then tick it off on the check list at the back of the book.

3 The collection of routine data

1 Why is routine data required?

To provide information on all aspects of economic life which can be used to prepare reports and make recommendations upon which management decisions can be taken.

2 What are primary statistics?

They are the raw data obtained directly as the result of an inquiry.

3 By contrast, what are secondary statistics?

These are obtained from some published source or from some quite different investigation which may nevertheless be useful to the inquiry we are currently pursuing. Examples of secondary data are the *Annual Abstract of Statistics* published by the UK Central Statistical Office.

4 What is meant by internal data?

These are data generated in house by routine reports or inquiries on such matters as sales, quality control, capital expenditure or specific problems such as labour turnover.

5 By contrast, what are external data?

These are data obtained from outside official or private sources such as the *Monthly Digest of Statistics, Economic Trends, Social Trends, The United Kingdom National Accounts* (Blue Book) on income and expenditure, *The United Kingdom Balance of Payments* (Pink Book), *The Family Expenditure Survey*, etc.
Note: Prepare a list of official sources of statistics available in your own town or country.

6 How are routine data collected?

Permanent records (such as names and addresses of customers, students, etc.) are collected when the opportunity first occurs, and are arranged alphabetically in a card index, loose-leaf or computer-based system. Current records, such as invoices and similar documents, returns, telemetered records, etc., are collated as they are produced, filed and treated statistically to give us the derived data we require.

7 What are 'returns'?

Regular reports called for by management from fieldworkers. They may be daily, weekly or monthly returns. They are called 'returns' because the forms are sent out by Head Office and have to be returned by given dates.

8 How are current records treated?

They can be used to prepare summaries, sub-totals and totals, either manually or using a computer system to extract relevant information.

9 How do regular returns aid management?

Regular returns provide management with information at regular, frequent intervals. By analysing these returns management can detect significant changes in behaviour or performance. For example, the ordinary register used in colleges and schools pinpoints students whose pattern of attendance is irregular. Management looks for the exceptional situation, the very good and the very bad and hence the system is called **management by exceptions**. Management might control a group of salesmen by comparing records on sales per month, sales calls made, kilometres covered, etc., and thus discover the strongest and weakest sales areas.

10 Why are computers useful for storing routine data?

In recent years low cost powerful computers have become available to all businesses. Data can be stored on magnetic tape, floppy disks or random access memories and displayed or printed as required.

11 Explain the individual items 1–6 in the computer configuration shown in Fig. 3.1 below.

(a) 1 is the microprocessor, which is the computer itself which stores, processes and manipulates the data fed into it.
(b) 2, 3 and 4 are all methods of feeding data into the microprocessor. 2 is the floppy disk drive unit. A floppy disk is inserted in the slot and its data are read into the computer. 3 is a cassette holder. Data stored on the tape of a casette can be read into the computer when the cassette is inserted into the casette holder. 4 is the keyboard, which can be used to insert new data as required. (c)

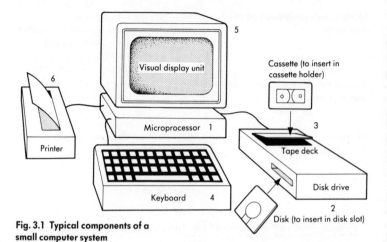

Fig. 3.1 Typical components of a small computer system

5 and 6 are methods of output from the computer. 5 is a VDU (visual display unit) which enables us to view the data inserted, and the results of the computer's data-processing. 6 is a printer which gives us a permanent record of the items viewed if we require it.

12 Given an example of such a system at work

Consider the sales records of a firm with many customers (debtors). The debtors file might be on a floppy disk, holding all the permanent records, names addresses and accounts. This would be inserted into the disk drive unit and read into the computer. Today's sales records could then be inserted by means of the keyboard. When these have been checked on the VDU they could then be processed – added to the appropriate accounts to update the balances on the accounts. They could then be re-stored on the disk. Once a month statements would be prepared for all accounts and sent off to the customers, requesting payment.

13 How can computers be used to analyse data?

Computers can handle large quantities of data rapidly and efficiently. Computer programs or software provide the instructions for a particular task to be performed and thus, provided that the software is available, any statistical analysis can be performed.

14 What is the computer doing in Table 3.1 below?

Monitoring certain data about the water quality of a river at times when it detects a significant change in one of the variables, and recording the whole set of data at that time.

Table 3.1 Routine data stored and listed by computer from river sensors

Date	Time	Dissolved oxygen (% saturation)	Temperature °C	Conductivity (μ siemens)	Ammonia concentration (mg/l)	pH
19.8–	09.09	50.00	15.60	947.21	0.04	7.73
19.8–	11.20	64.06	15.79	966.74	0.04	7.83
19.8–	11.49	66.79	15.79	966.74	0.04	7.86
19.8–	12.16	71.48	15.79	956.97	0.04	7.88
19.8–	12.19	72.26	15.89	956.97	0.04	7.89
19.8–	12.24	73.43	15.89	956.97	0.04	7.89
19.8–	12.34	75.78	15.89	956.97	0.04	7.93
19.8–	12.39	76.95	15.89	956.13	0.04	7.83
19.8–	12.44	76.56	15.89	956.97	0.04	7.99
19.8–	12.49	77.73	15.89	956.97	0.04	7.95

15 What are derived statistics?

They are further statistics derived from a mass of raw data. Examples are averages, percentages, standard deviations, etc.
Students should now consider the derived statistics provided by Table 3.2, which are the result of analysing water data obtained from river sensors.

Table 3.2 A statistical summary of routine data

Instrument	No. of samples	Mini-mum	Maxi-mum	Mean	Standard devia-tion	Unit
Dissolved oxygen	7	48.8	105.1	75.9	22.70	% saturation
Temperature	7	14.0	15.5	14.6	0.51	Celsius
Conductivity	7	869.1	957.0	927.7	33.35	μ siemens
Ammonia (as N)	7	0.04	0.04	0.04	0.00	mg/l
pH	7	7.7	8.3	8.0	0.21	—

Written Exercise: *Devise a 'returns' form for use in a youth club or evening institute to return statistics to the education office about attendance at classes. The form requires spaces to record 10 classes, giving five weeks attendances (some months have five weeks), the weeks to commence on Sundays. Instructions at the top require numbers in attendance to be in blue or black ink, except where attendance falls below 10, when they are to be completed in red. Fill in some imaginary classes and their attendances.*

Go over the topic again until you are sure of all the answers. Then tick it off on the check list at the back of the book.

4 Methods of statistical inquiry

1 In an inquiry what are the five chief ways of collecting data?

(a) Observation. (b) Inspection. (c) Abstraction from records. (d) Written questionnaire. (e) Interview.

2 Explain the method of observation

The situation under investigation is monitored unobtrusively by trained observers who record key points of interest, e.g. traffic using a particular road in a traffic census. Observers must be reliable and unbiased in their approach if valid results are to be obtained.

3 When is inspection used?

When manufactured objects or natural products are being tested for quality. Sampling is necessary and strict procedures must be adopted so that 'awkward' results are not disregarded. The inspectors should be independent if possible, with no vested interest in the results.

4 How does abstracting data help?

Past records may contain significant information relevant to the inquiry. It may be possible to use abstracted data to help make decisions and give us some insight into the problem.

5 What are the advantages of inquiries by written questionnaires?

(a) They reach the right people to respond – they are put into the in-trays of the person most likely to deal with them. (b) They can be carefully thought out and cover the subject of the enquiry comprehensively. (c) The cost is low, and expensive interviewers need not be employed. (d) The respondent doesn't have to give answers on the spur of the moment but may consult records, etc. (e) Responses do not all arrive at the same time and can be dealt with in turn as received.

6 What are the problems with written questionnaires?

They are a cheap and administratively convenient method of collecting data but may produce a low response rate with biased, irrelevant or 'joke' answers.

7 List the points to be remembered in drawing up a questionnaire

(a) Keep the instructions clear and simple, and as short as possible. (b) Make it clear who should answer it for preference, e.g. the transport manager. (c) If the completion of the form is compulsory, make this clear and tell the reader what the penalty is for non-compliance. (d) If it is voluntary, solicit the cooperation of the respondent and stress the non-personal nature of the inquiry and the preservation of privacy, if this is likely to be important. (e) Keep each question short – answerable by ticking yes/no boxes if possible. (f) If this is not possible, limit the answers; i.e., was it excellent, good, fair, poor or bad (tick the correct answer)? (g) Give clear instructions about the return of the questionnaire, address, post code, etc.

8 What is a 'schedule' as distinct from a questionnaire?

A schedule is a very similar list of questions which is taken round by an interviewer.

9 What are the advantages of interviewing?

(a) The interviewer is trained and soon becomes familiar with the schedule of questions. (b) Answering can proceed more rapidly and any misunderstanding can be explained. (c) A more subtle set of responses may result – if the interviewer has time to record them. (d) The interview can be evaluated later by the interviewer.

10 Why must interviewers be well trained?

To ensure that a systematic approach is adopted and to clear up any misunderstandings of the schedule of questions. The interviewer can then deal with any problems that arise as a real interview proceeds. The interviewer takes the respondent through a series of questions and at the end may give his/her impression of the interview and the veracity of the replies.

Written Exercise: *Design a questionnaire to be used to keep a record of accidents in a small factory. The forms will be sent to Head Office to acquaint them with all the particulars of each accident, date, time, nature of the occurrence, who was involved, who treated the injured person, etc. A space for signature of the supervisor concerned should be provided.*

Go over the topic again until you are sure of all the answers. Then tick it off on the check list at the back of the book.

5 Censuses and samples

1 What is a census?

A statistical inquiry in which every member of the population is investigated.

2 What does the word 'population' mean?

It can mean the entire citizens of a nation state – but in statistics it means the whole set of people affected by or interested in the problem under examination. .

3 When is a national census held in the United Kingdom?

Every ten years, in the first year of a decade, i.e., 1971, 1981, etc.

4 What does it reveal?

It reveals information on population levels, age distribution, housing needs, social service requirements, etc.

5 Why is a census often impracticable in an enquiry?

It would cost too much to investigate the whole population. We can get almost as reliable information from a random sample of the population.

6 What is a random sample?

A sample selected in such a way that every member of the population has an equal chance of being selected.

7 How can we select a random sample?

(a) By numbering the population and drawing the numbers at random as in a Bingo game *or* (b) By generating them in a computer.

8 How can I do this on a computer?

Look at your own or your firm's computer handbook. You will probably find a random-number generator function called RAND(Y). Try using it to generate your own random number series. An example is shown in Table 5.1.

Table 5.1 Random numbers

11	74	21	82	26	29
00	43	40	72	61	19
68	02	40	44	17	75
11	71	08	72	33	93
23	05	79	82	64	57
31	06	06	51	19	26
48	64	73	53	43	36
19	94	36	42	75	34
55	71	15	47	44	91

9 What are the other methods of sampling?

(a) Systematic (periodic or equal interval) sampling. (b) Stratified sampling. (c) Multi-stage or area sampling. (d) Quota sampling. (e) Attribute sampling.

10 Explain a systematic (periodic or equal-interval) sample

A systematic sample is one which selects a starting point by random methods and then selects items systematically from this starting point, i.e., 7th, 57th, 107th, etc.

11 Explain stratified sampling.

Stratified sampling selects a representative sample by dividing the population into strata, and ensuring that a fair proportion of items from each stratum appears. For example, an inquiry about the opinions of car owners should not just consult the owners of Rolls Royces or the owners of minis. We should divide the cars into ranges, assess the number in each range and ask a fair proportion of drivers in each range for their views.

12 Explain multi-stage sampling

Multi-stage sampling selects randomly subareas of a total population and then selects individuals to be interviewed by random selection from the subareas. Thus a population survey might select counties by random selection, then a

15

few local authority areas within the counties selected and then individuals in these local authority areas.

13 Explain quota sampling

In quota sampling we break down the population to be investigated into strata and then set a quota (the number of individuals to be approached) for each band. We make no attempt to find actual named individuals. Thus we might require the interviewer to direct the questions at 4 mechanics, 2 teachers, 5 housewives, a lorry driver, etc.

14 Explain attribute sampling

We select individuals displaying particular attributes, for example students aged 17, or physically handicapped, or those born in April.

Written Exercise: *Distinguish between a census and a sample. What are the advantages and disadvantages of sampling?*

Go over the topic again until you are sure of all the answers. Then tick it off on the check list at the back of the book.

6 Classification and tabulation of data

1 What is the result of any inquiry, whether by observation, questionnaire or interviewing?

A mass of raw data.

2 Why is raw data first screened?

Because any mass of raw data may include incorrect answers or even 'joke' answers. It is necessary to screen the data to remove these manifest absurdities and classify the data to reduce it to manageable proportions.

3 What is the essential feature of classification?

To group items together naturally into categories which together are comprehensive and therefore able to embrace all the data. Categories should not be too numerous, should not overlap and each category should be homogenous (i.e., each category should display some common characteristics (e.g., hatch-backs, saloons and estate cars in a survey of cars).

4 How is the data rearranged into an array?

The data is sorted in increasing order of size according to some particular variable or attribute such as weight or value. Thus for an inquiry on monthly earnings the array starts with the smallest monthly pay packet (say £128) and finishes with the largest (perhaps £925).

5 Explain how to prepare a frequency distribution

An array is in increasing order of size but does not show the frequency with which each individual statistic occurs. A frequency distribution brings together similar statistics and shows the number of times they occur. Table 6.1 shows a typical frequency distribution.

Table 6.1 Frequency distribution of average monthly take-home pay: year 19—
(Correct to nearest £1 sterling)

Value (£)	Frequency	Value (£)	Frequency	Value (£)	Frequency	Value (£)	Frequency
128	1	160	1	168	2	184	3
129	2	162	1	171	1	185	1
154	1	164	2	172	1	186	3
157	1	165	1	176	1	188	2
158	1	166	1	179	1	192	1

6 What is a grouped frequency distribution?

It is a table which further reduces the spread of an array by grouping the variables into a smaller number of specified classes, or bands of value, and giving the frequencies in each group. An example is given in Table 6.2.

Table 6.2 Grouped frequency distribution of average monthly take-home pay: year 19—
(Correct to nearest £1 sterling)

(£)	Frequency
100 but under 200	33
200 but under 300	42
300 but under 400	31
400 but under 500	20
500 but under 600	3
600 but under 700	5
700 but under 800	12
800 but under 900	3
900 but under 1000	1
Total	150

7 How do we produce a grouped frequency distribution?

The full range of data is divided into a number of subgroups, or classes, using subdivisions which are as natural as possible. The number of items which fall into each subgroup is calculated and tabulated as shown in Table 6.2.

8 What is a cumulative frequency column?

It is an extra column added to a grouped frequency distribution which cumulates the frequencies from start to finish (or they can be accumulated in reverse order, from finish to start).

9 Calculate the cumulative frequencies and the reverse cumulative frequencies for Table 6.2.

Table 6.3 Cumulative columns on a grouped frequency distribution of average monthly take-home pay: year 19—.
(Correct to nearest £1 sterling)

Take-home pay (£)	Frequency	Cumulative frequency	Reversed cumulative frequency
100 but under 200	33	33	150
200 but under 300	42	75	117
300 but under 400	31	106	75
400 but under 500	20	126	44
500 but under 600	3	129	24
600 but under 700	5	134	21
700 but under 800	12	146	16
800 but under 900	3	149	4
900 but under 1000	1	150	1
Total	150		

10 What sort of statements can we make from these cumulative frequencies

We can make statements such as: (a) 126 of the 150 employees earned less than £500 per month. (b) Only 16 of the 150 employees earned £700 or more per month.

11 What is tabulation?

It is the process of arranging data in a table, such as a grouped frequency distribution, for final presentation in a report.

12 What are the essential requirements of tabulation?

Tables must be clearly titled with a source reference (if any) given. Classes should be natural and totals and other derived statistics included if necessary to bring out the salient points of the enquiry. Rounding to eliminate excessive detail may be helpful.

13 How do tables help us understand problems?

They are made up of relatively straightforward data, presented to display information so that we can understand it at a glance.

14 What is meant by rounding?

Data often contains more figures than can be comprehended by most people and it is better to round the figures to simplify them. We would round the population of a country from 55 295 321 to 55 million, for example.

15 What are the rules for rounding?

We look at the part of the number to be discarded and ask ourselves. Is this more than half or less than half? If it is less than half the smallest place value to be shown – for example millions – we discard it. If it is more than half, we count it as an extra million. If it is exactly half, for example 500 000 – we round it to the nearest even million. So 55 500 000 becomes 56 million, but 54 500 000 becomes 54 million.

Written Exercise: *Serious crimes in a certain year include 132 murders, 265 cases of manslaughter, 7500 cases of wounding, 1854 offences against the person, 7204 charges of sexual nature, 38 750 burglaries, 2999 burglaries and 20 406 cases of handling stolen goods. Fraud and forgery total 17 705. List these crimes in a table, to the nearest 100, and bring out the total. Now calculate each group as a percentage of the whole, using your abbreviated figures.*

Go over the topic again until you are sure of all the answers. Then tick it off on the check list at the back of the book.

7 Approximations and errors

1 Why is it necessary to check for errors?

All data contain errors. Sampling and measurement errors are inevitable, and the statistician must be able to discriminate between 'good' and 'bad' data and eliminate mistakes wherever possible.

2 What errors occur at the interview or questionnaire stage?

The interviewer may record a wrong answer or misinterpret the answer given. The respondent to a questionnaire may misunderstand the question.

3 What errors occur at the classification stage?

There may be mistakes transferring answers to cards on computer input. There may be errors in tabulation and frequency analysis. The choice of classes may be poor, and errors may be made by incorrectly classifying the answers.

4 What errors occur at the evaluation stage?

Mistakes may be made in the calculation of derived statistics and table headings may be ambiguous and misunderstood.

5 How does rounding introduce errors?

By approximating data and rounding to the nearest whole or reasonable number, errors are introduced. Such errors may be multiplied or accumulated during statistical analysis and may yield inaccurate answers to management questions.

6 What is meant by bias?

Bias is the tendency of statistics to be influenced in a particular direction and thus give an incorrect impression. For example, in preparing cost forecasts departmental managers may exaggerate likely future costs; in this case the costs are biased upwards.

7 How can rounding introduce bias?

If numbers are rounded up or down, to, say, the nearest thousand, upward or downward bias is introduced. Normal rounding produces unbiased errors in that errors cancel one another out sooner or later.

8 Remind us again of the rules for unbiased rounding

If the portion of the data being disregarded is more than half, round up (e.g. 27 675 is rounded up to 28 000).

If the portion is less than half, round down (e.g. 25 321 goes to 25 000). If the portion is exactly half round to the nearest even thousand (e.g. 27 500 goes to 28 000 and 26 500 goes to 26 000).

9 Explain the concept 'absolute error'

This is the difference between the approximate figure and the original quantity (e.g. 28 000 minus 27 675 equals 325). The rounded figure is +325 compared with the actual figure.

10 What is a relative error?

This is obtained by considering the absolute error as a fraction of the rounded total (e.g. $\dfrac{325}{28\ 000}$ gives a relative error of 0.0116 or 1.16% in percentage terms).

11 How do we calculate the absolute error of a sum of rounded numbers?

By summing the absolute errors in the original numbers. For example, 9000 (to the nearest hundred) added to 500 (to the nearest ten) gives 9000 ($\pm$50) + 500 ($\pm$5) which equals 9500 ($\pm$55).

12 What happens if we are subtracting rounded numbers?

The absolute error of the difference equals the sum of the absolute errors in the original numbers. For example 93 000 (to the nearest 1000) less 6600 (to the nearest 100) gives 93 000 ($\pm$500) − 6600 ($\pm$50) which equals 86 400 ($\pm$550).

13 Explain the rule for multiplication

The relative error of a product is the sum of the **relative** errors in the multiplier and the multiplicand. Thus 9000 (to the nearest hundred) multiplied by 500 (to the nearest ten) gives (9000 $\pm$ 0.56%) (500 $\pm$ 1.0%) which equals
$$4\ 500\ 000 \pm 1.56\%$$
$$= 4\ 500\ 000 \pm 70\ 200.$$

14 And for division?

The relative error of the quotient is the *algebraic difference* of the relative errors in the dividend and the divisor (the maximum error is therefore the same as the sum of the relative errors). Thus 9000 (to the nearest hundred) divided by 500 (to the nearest 10) gives $(9000 \pm 0.56\%) \div (500 \pm 1\%)$ which equals $18 \pm 1.56\% = 18 \pm 0.2808$.

Written Exercise: *What types of error might result from conducting an inquiry into eating habits by a written questionnaire?*

Go over the topic again until you are sure of all the answers. Then tick it off on the check list at the back of the book.

8 The pictorial representation of data: charts and diagrams

1 What are the chief types of pictorial representation of data?

(a) Graphs. (b) Pictograms. (c) Bar-charts. (d) Pie charts. (e) Gantt charts. (f) Histograms.

2 Why are graphs so important?

Data is frequently displayed as graphs and these allow rapid assimilation of information. They gave rise to the whole art of graphics. (Topic 9 is devoted to them).

3 What are pictograms?

These convey statistical facts in picture form. An appropriate symbol is used to represent the data, and the size or number of symbols denotes the quantity involved as shown in Fig. 8.1.

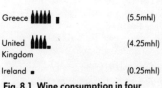

Greece (5.5mhl)

United Kingdom (4.25mhl)

Ireland (0.25mhl)

Fig. 8.1 Wine consumption in four countries (Source: *Annual Reports: EEC*)

Italy (72.25mhl)

4 What does each symbol (a bottle) display in Fig. 8.1?

One million hectolitres.

5 Explain the principle of the bar chart

Information is related to the horizontal or vertical length of a bar or thick line. The bars are used to compare sets of data. This is shown in Fig. 8.2 in an analysis of turnover in a supermarket.

Pounds sterling

	0	250	500	750	1000	1250	1500	1750
Groceries	£1750							
Household items	£1000							
Greengroceries	£875							
Cigarettes and tobacco	£750							
Wines and spirits	£375							

Fig. 8.2 An analysis of turnover in a supermarket (Source: Week 17 Departmental Returns)

6 What are the design features of a bar chart?

(a) The *scale* must be chosen so that all the data can appear easily, i.e. with the largest piece of data using up almost all the scale. (b) The bars should be the same width, since only length is used to denote the data. (c) Bar charts can be drawn horizontally or vertically. (d) Colours or cross-hatching can be helpful in differentiating between the bars. (e) Positive and negative bar charts can be used to show relative changes about a mean or zero, as shown in Fig. 8.3.

7 When are multiple bar charts used?

When we wish to compare a number of items over a number of years. They show how each item varies over the period. (An example is given in Fig. 8.4.)

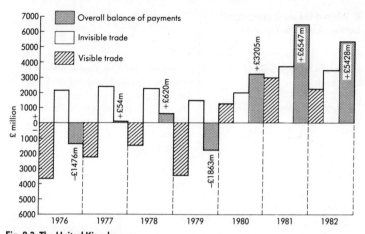

Fig. 8.3 The United Kingdom
balance of payments 1976–82
(Source: *The United Kingdom
Balance of Payments* (Pink Book))

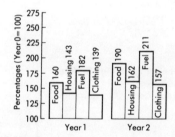

Fig. 8.4 Multiple bar charts for
comparing data

**8 How would you show totals
using a bar chart?**

By drawing a component bar chart as
shown in Fig. 8.5.

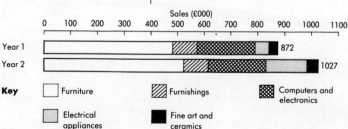

Fig. 8.5 Component bar charts

9 Explain Fig. 8.5

The total sales in each year is shown by the total length of the bar (£872 000 in Year 1 and £1027 000 in Year 2). The individual parts that make up the total are shown by shading, cross-hatching, etc., and can be compared.

10 Describe a percentage bar chart?

This is obtained by showing the entire set of statistics as 100 per cent and components calculated as a percentage of the whole (see Fig. 8.6).

100% = £342m

58.8%	To employers (£201m)
12.0%	To governments (£41m)
9.6%	To investors (£33m)
19.6%	To expansion of business (£67m)

Fig. 8.6 A percentage bar chart on the use of profits

11 Do three-dimensional bar chart representations convey any additional information?

Very rarely, and they make it more difficult to see the end of the bar. They may add to the visual impression, but are awkward to draw quickly in examinations, and should be avoided. An example is given in Fig. 8.7.

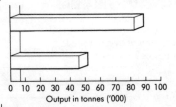

Fig. 8.7 A three-dimensional bar chart

0 10 20 30 40 50 60 70 80 90 100
Output in tonnes ('000)

12 What is a pie chart?

As shown in Fig. 8.8, the round 'pie' is cut up into slices, each of which represents a percentage of the total. The percentage is translated into degrees so that the total is 360°. These degrees have been given in Fig. 8.8 to show how the circle is divided up, but they would usually not appear in the finished pie chart. Instead the percentages of the total would be shown.

UK visible exports

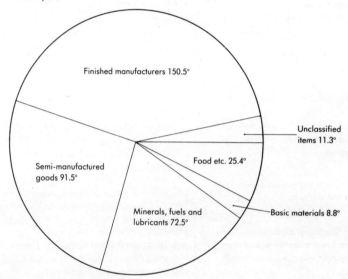

Fig. 8.8 United Kingdom visible exports (Source: *The United Kingdom Balance of Payments* (Pink Book))

13 What chart would you use to compare actual and planned performance?

The Gantt chart.

14 How do we display them on a Gantt chart?

The chart consists of a layout of equally spaced columns, each of which represents a week (or month). Planned performance is represented by the full width of the column, and actual performance is filled in across the column as it is achieved. If the other side

27

of the column is not reached we have under-performed, and if performance is greater than planned we start a second line. This is drawn in Fig. 8.9.

	1	2	3	4	5	6
Planned sales (£)	4000	4500	3250	4000	5000	8000
Actual Sales (£)	3400	4250	3800	4750	6250	
% achievement	85%	94.4%	116.9%	118.75%	125%	
Actual achievement						
Cumulative achievement						

Fig. 8.9 A Gantt chart

15 Explain the cumulative achievement line on the Gantt chart

The cumulative achievement line is built up by adding each period's performance to the previous performances. We do have to be a little careful though. The whole point of a Gantt chart is that it is flexible – the planned performances for each period do not have to be the same. For example, sales figures could reflect seasonal changes. We might plan for massive sales in December and very little in January, after the Christmas rush. In building up the cumulative line we have to show the last section of actual achievement as a percentage of the planned performance of that period. In Fig. 8.9 the extra performance achieved in the five months is £1700 sales, which is less than $\frac{1}{4}$ of £8000 – the planned sales in the next period.

16 What is a histogram?

A way of displaying frequency distributions so that they can be easily understood.

17 How is a histogram drawn?

As a series of rectangles each of which represents one class interval. The blocks stand next to one another and show the pattern of the distribution. If the blocks are of uniform width the height of each block is determined by the frequency, as in Fig. 8.10 below. If not, the height must be adjusted, because in a histogram it is the area of the block that represents the data.

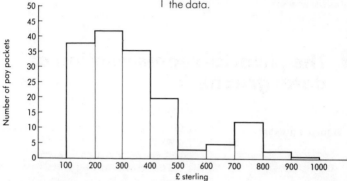

Fig. 8.10 A histogram on monthly take home pay

17 What is a frequency polygon?

It is a line drawn on the histogram by joining the midpoints of the tops of the blocks as shown in Fig. 8.11. Each line creates two similar triangles, one of which is included in the polygon and the other excluded. The result is that the area under the polygon is the same as the total area of the blocks.

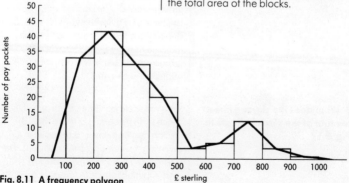

Fig. 8.11 A frequency polygon

Written Exercise: *A United Kingdom firm sells 46 per cent of its goods to Europe, 13 per cent to the USA, 18 per cent to African countries, 12 per cent to India, 7 per cent to Pakistan and the balance to Australia. Draw a pie chart to show its distribution to overseas territories.*

Go over the topic again until you are sure of all the answers. Then tick it off on the check list at the back of the book.

9 The pictorial representation of data: graphs

1 What are graphs?

Graphs are pictorial representations of data which show the relationship between two variables. The variables are plotted against axes called the *x* and *y* ordinates. A scale is chosen to match the data and crosses or some other symbol are used to mark the points on the graph. The points may be joined by a continuous or dotted line or a combination of both if more than one set of data is to be drawn.

2 Explain Fig. 9.1

Figure 9.1 shows sales of newspapers (shown as the *y* ordinate) plotted against quarters of the year (shown as the *x* ordinate). The sales of each paper are joined by different types of lines to differentiate them for the reader. They indicate that the *Daily Pictorial* has overtaken the *Daily News* in the three-year period shown.

3 What does the zig-zag line at the foot of the *y* axis indicate in Fig. 9.1?

That the scale has been interrupted. Since both papers sold more than 3 million copies there was no point in showing the sales below that figure. The vertical scale is interrupted between 0 and 3 to highlight data variations.

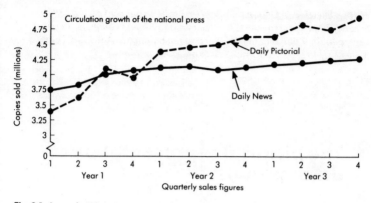

Fig. 9.1 A graph with an interrupted scale

4 What is Fig. 9.2? | It is a Z chart.

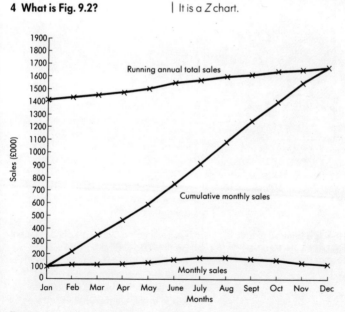

Fig. 9.2

5 Explain what a Z chart is

It is frequently used by selling organisations to plot current, cumulative and annual figures. The result finishes up as a rather erratically shaped Z curve, as shown in Fig. 9.2. The bottom line shows current monthly sales. The top line shows a moving cumulative total of sales for the previous 12 months. The oblique line shows the cumulative total for the current year to date. The Z must join up in the last month of the year because the total for the year to date then becomes the same as the total for the last 12 months.

6 What type of graph is Fig. 9.3?

A layered graph.

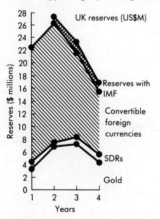

Fig. 9.3 (Source: *The United Kingdom Balance of Payments* (Pink Book))

7 How is a layered graph obtained?

Quite often data consists of component parts of a grand total. Total sales of a firm would be made up of sales from several departments or the official reserves of the United Kingdom are made up of several types of finances, as shown in Fig. 9.3. As gold is heavy, we show it as the bottom layer. The second layer – IMF special drawing rights – is then added and further layers are superimposed until the total United Kingdom reserves have been obtained as the top line on the graph.

8 What sort of graph is Fig. 9.4?

A straight-line graph through the origin.

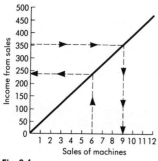

Fig. 9.4

9 What does this type of graph show?

It shows a constant relationship between two variables. From such a graph we can interpolate (find the value of any point on the graph). For example we can read off the income from any volume of sales by simply drawing a vertical line from the horizontal axis to intercept the graph and then drawing a horizontal line from the intercept point to the *y* axis.

10 What is extrapolation?

Reading off the value of a point not on the graph as drawn, but on an extension of the graph, to predict the future sales, etc.

11 Why is extrapolation sometimes misleading?

Because when we extrapolate outside the range of the graph we cannot be sure that the straight-line relationship still holds. It does not follow that because 10 machines can be sold for £400 we can earn £400 000 by selling 10 000 of them. We may have to lower the selling price to achieve the larger volume of sales.

$y = a + bx$

12 What is the general expression for a straight line graph showing a relationship between two variables x and y?

13 Will such a graph pass through the origin?

Only if a, the constant term, is 0.

14 What does a decide in the formula $y = a + bx$?

The intercept on the y axis, so if a is zero, the intercept will be at the origin of the graph.

15 What does b decide in the formula $y = a + bx$?

The slope of the graph. It is the extent to which y changes with respect to x.

16 What sort of graph is Fig. 9.5 below?

A breakeven chart

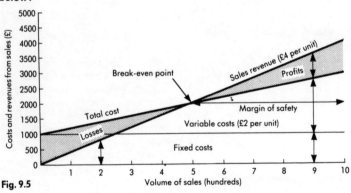

Fig. 9.5

17 What purpose do breakeven charts serve?

They provide a means of assessing the effect of pricing policy and fixed variable costs on profitability. Figure 9.5 shows the total costs made up of a fixed costs (e.g., building costs) plus variable costs (e.g., operating costs) and the total revenue obtained from sales. The point where the two lines cross is called the breakeven point where costs and revenue are equal. Operating to the right of this point gives us an overall profit – to the left produces a net loss.

18 How can we emphasise the relative importance of two variables?

By rewriting the data in cumulative percentage terms and then plotting the new data set for the variables against each other, we obtain a Lorenz curve.

19 Explain the Lorenz curve in Fig. 9.6 which is about the distribution of incomes in Espaniona

(a) The straight line is the locus of points where income is being equally distributed. (b) Since our data points are far from this line they indicate that income is not evenly distributed (only 37 per cent of income is earned by 94 per cent of the population. (c) Espaniona is clearly a country where the rich are very rich and many people are rather poor. (50 per cent of the population have about 14 per cent of the income.)

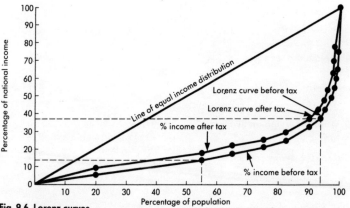

Fig. 9.6 Lorenz curves

Written Exercise: A firm has fixed costs of £20 000 and variable costs of £10 per unit of output. Each unit is sold for £20. Draw a graph of output and sales in steps of 1000 up to 5000 units. Where is the breakeven point?

Go over the topic again until you are sure of all the answers. Then tick it off on the check list at the back of the book.

10 Measures of central tendency – 1: the mean

1 What is a measure of central tendency?

It is an average which can be used as a starting point to describe a distribution. The average is a point within a group of data which is central to the group and around which the other values are distributed.

2 List four types of average

(a) The arithmetic average (or arithmetic mean). (b) The geometric average (or geometric mean). (c) The median (or central item in an array). (d) The mode (or most fashionable item).

3 How is the arithmetic mean (a.m.) calculated?

It is found by adding together individual values and dividing by the number of items. The formula for a simple arithmetic mean is

$$\bar{x} = \frac{\Sigma x}{n}$$

4 Explain this formula

$\bar{x}$ stands for the arithmetic mean, Σ means 'the sum of', x is used for the individual items and $n =$ the number of items.

5 What is a weighted arithmetic mean?

Where a particular item occurs a number of times it is said to weight the statistics in its favour. Consider the data shown in Table 10.1. To calculate the arithmetic mean, we must multiply each rate of pay by the number of employees receiving that rate, to give a total pay at each rate. We then divide the grand total by the sum of the frequencies (i.e., the total number of employees).

Table 10.1 Wages of Employees

Weekly wage (£)	Number of employees	Total pay (£)
85	15	1 275
90	53	4 770
95	25	2 375
100	50	5 000
120	44	5 280
Total	187	18 700

6 Give this as a formula for a weighted arithmetic mean

$\bar{x} = \frac{\Sigma fx}{\Sigma f}$. Here f is the frequency (the number receiving each rate of pay). As $\Sigma f = n$ (the number of items) we could have said $\bar{x} = \frac{\Sigma fx}{n}$

7 What is the average wage of employees in Table 10.1?

$\frac{£18\ 700}{187} = £100.00$

8 How do we simplify the a.m. calculation for large numbers?

We work from an assumed mean which is chosen to be, for example, the rounded lowest value in a set of data. The true mean will be above this and to find it we average the residue of the numbers when the assumed mean has been deducted from them. Consider the numbers 23 874, 23 525, 23 004, 24 156 and 23 126. If we take our assumed mean to be 23 000 the residuals are 874, 525, 4, 1156 and 126. The sum of these is 2685 and dividing by 5 gives the average residual as 537. Therefore the true mean is 537 above 23 000, which is 23 537.

9 Table 10.2 shows grouped distribution data. Calculate the a.m. for this data set

Table 10.2 Output of operatives: weeks 1–4 inclusive

Output in units	Midpoints	Number of operatives	Products
21–30	25.5	7	178.5
31–40	35.5	11	390.5
41–50	45.5	14	637.0
51–60	55.5	8	444.0
61–70	65.5	5	327.5
Total		45	1 977.5

We assume that the data is evenly distributed throughout each group and multiply the group midpoints by the frequency (number of operatives). The sum of these products, 1977.5, is then divided by the sum of the frequencies (i.e., the total number of operatives) to give $1977.5 \div 45 = 43.94$. This is the a.m. of the distribution.

10 How do we deal with open-ended distributions?

We have to make an assumption about the size of the open-ended classes. We usually assume that the end group is twice as large as other groups.

11 What is the formula then for finding the arithmetic mean of a grouped frequency distribution?

$\bar{x} = \dfrac{\Sigma fx}{\Sigma f}$ (as before for a weighted average – but in this case x stands for the midpoints of the groups, not the individual items. f is the frequency of each group).

12 If the numbers are very large in each of the groups, the multiplication to get the products becomes tedious – even with a calculator. What can we do to reduce the work?

Use the short-cut method of starting from an assumed mean.

13 How does this work?

Table 10.3 Output of operatives: weeks 1–4 inclusive

Output in units	Midpoints	Number of operatives	Deviation from assumed mean (45.5) in class intervals (CI)	Products
21–30	25.5	7	−2	−14
31–40	35.5	11	−1	−11
41–50	45.5	14	0	0
51–60	55.5	8	1	8
61–70	65.5	5	2	10
Total		45		− 7CI

If we assume the mean is at the midpoint of the 41–50 group in Table 10.3, i.e. it is at 45.5, we determine the deviation of the groups from the assumed mean *in terms of class intervals*. We then multiply these deviations by the frequency in each group to give a series of products. The sum of these products (−7) is divided by the sum of the frequencies (45) to give −0.156 class intervals. Since the class interval is 10 the deviation is −1.56 and the a.m. is $45.5 − 1.56 = 43.94$.

14 What if the groups are of different sizes?

We overcome this difficulty by using a standard class interval in the calculation. Suppose that in Table 10.3 the last group had been 61–90, the midpoint would then have been 75.5, and the deviation from the assumed mean would have been 3 standard class intervals. The calculations would have been adjusted accordingly.

15 Express the 'assumed mean' method in a mathematical formula

$$\bar{x} = x' + \left(\frac{\Sigma fd'}{\Sigma f} \times i \right)$$

where x' means the assumed mean, d' means the deviations from the assumed mean in class intervals and i is the class interval.

Written Exercise: *Students admitted to university for undergraduate courses were found to have the following ages:*

Age	Males	Females
16	1	4
17	2	—
18	87	75
19	96	85
20	54	65
21	42	51
22	17	19
35	1	—
55	—	1
Total	300	300

a *Find the arithmetic mean age for males and females separately.*

b *How may these means be combined to give the mean age for all students?*

Go over the topic again until you are sure of all the answers. Then tick it off on the check list at the back of the book.

11 Measures of central tendency – 2: the median

1 What is a median?

It is the value of the central item in an array; i.e., the central item in a set of data arranged in increasing order of size.

2 Suppose you are given the set of data 27, 29, 13, 14, 12, 48, 37. What would you do to find the median?

Rearrange it as an array 12, 13, 14, 27, 29, 37, 48. The median is the value of the central item, 27.

3 What is the formula for finding the median in a simple array?

$\frac{n+1}{2}$ where n is the number of items.

There are 7 items in the array above so $\frac{7+1}{2} = \frac{8}{2} = 4$. The median is the value of the 4th item, which is 27.

4 'Suppose there are an even number of items as with this set of data: 12, 13, 14, 27, 29, 37, 48, 57

The median $\frac{n+1}{2} = \frac{8+1}{2} = 4\frac{1}{2}$.

The median is the $4\frac{1}{2}$th item – so we have to take the average of the 4th and 5th items. This is $\frac{27+29}{2} = \frac{56}{2} = 28$.

5 How do we find the median of grouped data?

We cannot find an exact median because we don't know how the data in any group are spread. We have to make an assumption that the data are evenly spread in the group. The formula for grouped data is $\frac{n}{2}$, not $\frac{n+1}{2}$. There is a technical explanation for this, but it need not worry us here.

6 All right then. What shall I do to find the median of the set of data below?

Output in units	Number of operatives
21–30	7
31–40	11
41–50	14
51–60	8
61–70	5

We set the data down with a cumulative frequency column alongside

21–30	7	7
31–40	11	18
41–50	14	32
51–60	8	40
61–70	5	45

There are 45 items, so the median is $\frac{n}{2} = \frac{45}{2} = 22\frac{1}{2}$th item. It must therefore be in the 41–50 group. It is the $4\frac{1}{2}$th item out of 14 in the group. We therefore find it by

$$40 + \left(\frac{4\frac{1}{2}}{14} \times 10\right) = 40 + \frac{90}{28}$$
$$= 40 + 3.21 = \underline{43.21}\ \text{units}$$

7 How else can we find the median?

By interpolation on a cumulative frequency curve.

8 How is this done?

We plot the data as a cumulative frequency curve and then read off the value of the median item. This has been done in Fig. 11.1, using the data in Question 6 above. (Continues overleaf)

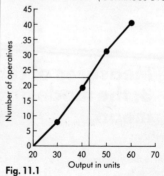
Fig. 11.1

41

The cumulative values are plotted against the upper limits of the groups of data, 7 against 30, 18 against 40, etc.

The median can then be read off as the $22\frac{1}{2}$th item = 43. This is as accurate as we can get with a graph.

9 What are the features of the median as an average?

(a) It is easy to find – the central item in the array. (b) It divides the data into two halves – half on one side, half on the other. (c) It takes absolutely no account of any other items; so extreme items do not distort the picture. (For example, if a class has 10 students aged 17 and one adult of 67 the median declares the average age to be 17. The mean says it is 21.55.) (d) Its disadvantage is that it is no use for mathematical manipulation. Thus the mean multiplied by the number of items gives the total value, but the median does not. (In the example above

$$21.55 \times 11 = 237.05$$
$$17 \quad \times 11 = 187$$

Written Exercise: *A player throwing a die scores 1, 5, 6, 2, 4, 1, 1, 2, 2, 4, 6, 5, 6, 3, 4, 2, 1, 5, 5, 6, 4, 1. What was the median score?*

Go over the topic again until you are sure of all the answers. Then tick it off on the check list at the back of the book.

12 Measures of central tendency – 3: the mode and the geometric mean

1 What is the mode of a distribution?

It is the value which occurs most frequently in the distribution.

2 Why is the mode particularly useful?

It provides information on the most frequent occurrence or most popular (fashionable) item. This can be a better method of describing a data set than the mean or the median.

3 Give examples

The modal number of legs for a human being is 2, but the mean would be somewhat less than 2 because some people have lost a leg in accidents, etc.

When houses are advertised, we usually specify the number of bedrooms. To say that the modal number of bedrooms is 3 is simpler to understand than saying that the mean number of bedrooms is 3.147 bedrooms per house.

4 What is meant by a bimodal and a multimodal series?

It is quite likely that in a distribution 2 or more items are equally popular. The series is bimodal if 2 occur with the same frequency and multimodal if more than 2 occur with the same frequency.

5 What are the problems with a grouped distribution?

We cannot tell exactly where the mode is, and consequently have to make assumptions.

6 What are they?

(a) That the mode lies in the group with the highest frequency. (b) That the frequencies in all the groups are evenly spread within the group.

7 Does the modal item lie at the midpoint of the group with the highest frequency?

No, because the pattern of the distribution influences our idea of where the mode should be. For example, consider the data set given below:

Number of units produced	Number of employees
50 and under 75	23
75 and under 100	45
100 and under 125	86
125 and under 150	11
150 and under 175	5
Total	170

The model group is the '100 and under 125' group, where 86 employees were achieving this level of output. As there are 45 people in the group below this and only 11 in the group above it we would feel that the mode must be rather nearer the beginning of the '100 and under 125' group than the end. To find where it is we must either draw a histogram or do a calculation.

8 Draw the histogram and hence find the mode.

Notes: (a) The mode is in the largest group, but displaced from the centre by the relative importance of the groups above and below it. (b) By joining up as shown we find the mode where the two lines intersect, at about 109 units. This is the modal output.

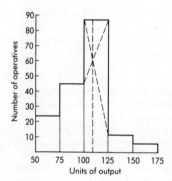

Fig. 12.1

9 What formula would we use to do the calculation?

$$\text{Mode} = L + \left(\frac{f_m - f_l}{(f_m - f_l) + (f_m - f_h)} \right) \times CI$$

where L is the lowest point of the modal group,
f_m is the frequency in the modal group,
f_l is the frequency in the lower group,
f_h is the frequency in the higher group,
and
CI is the class interval.

10 Do the calculation for the data in 7 above

Mode

$$= 100 + \left(\frac{86 - 45}{(86 - 45) + (86 - 11)} \right) \times CI$$

$$= 100 + \left(\frac{41}{41 + 75} \right) \times CI$$

$$= 100 + \frac{41}{116} \times 25$$

$$= 100 + 8.836$$

$$= \underline{108.8 \text{ units}}$$

11 What is the geometric mean?

It is an average found by multiplying n variables together and then taking the nth root of them. Thus the geometric mean of 4 and 7 is the square root of 4×7

$$\text{G.M.} = \sqrt{28} = \underline{5.29}$$

12 Give the general formula for a geometric mean (G.M.)?

$$\text{G.M.} = \sqrt[n]{x_1 \times x_2 \times x_3 \ldots x_n}$$

so the geometric mean of 3, 25 and 45. would be

$$\sqrt[3]{3 \times 25 \times 45} = \sqrt[3]{3375} = 15$$

13 When is the geometric mean used?

It is useful when only a few items in a distribution are changing since it is more stable than the arithmetic mean. The *Financial Times* Industry Ordinary Share Index is calculated as a geometric mean. It also provides a better description of population data, since population tends to increase in a geometric progression.

Written Exercise: *Draw a histogram to find the modal output of technicians whose results are as follows: less than 20 units, 14 operatives; 20–29 units, 36 operatives; 30–39 units, 84 operatives; 40–49 units 46 operatives; 50 or more units, 11 operatives.*

Go over the topic again until you are sure of all the answers. Then tick it off on the check list at the back of the book.

13 Measures of dispersion – 1: the range and the quartile deviation

1 Explain the term 'dispersion'

Dispersion is a measure of the extent to which the individual items of a set of data are spread around the average.

2 What is meant by a normal distribution curve?

Here the individual observations are grouped around the midpoint in a bell-shaped symmetrical curve, as shown in Fig. 13.1, about the tossing of a coin.

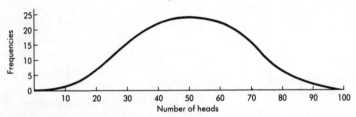

Fig. 13.1 The normal distribution curve

3 How do these normal curves vary?

The distributions around the central point may differ, as shown in Fig. 13.2, where in curve ss the average deviation from the mean is larger than with the curve tt. The more closely the data are grouped around the mean, the higher the peak of the graph, for the frequencies in the narrow band around the mean must be greater.

4 Where are the mean, median and mode located in the normal curve?

They are all exactly the same and positioned centrally. By definition, the mode is always found at the apex of the curve. The median which is the value of the central item will always be at the apex of a normal curve, too, and so will the mean.

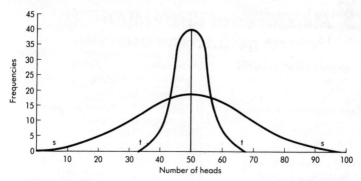

Fig. 13.2 Normal curves with different distributions

5 When is a frequency distribution said to be skewed?

When it is not symmetrical. Figure 13.3 shows two skewed curves. If the majority of the data lie to the left of the mean position, the distribution is positively skewed (curve AAA). Curve BBB is negatively skewed with the majority of the data to the right of the mean position. Notice the changed positions of the mean and median relative to the mode.

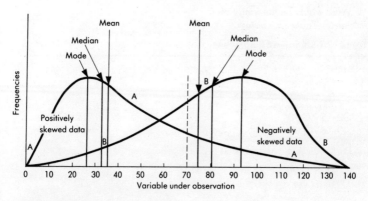

Fig. 13.3 Positively skewed and negatively skewed distributions

6 Give some examples of positively and negatively skewed distributions

National wage structures tend to be positively skewed, since most people earn relatively law wages but a few really high income earners drag the average income up disproportionately, away from the modal position where the mass of incomes is situated. Data related to health are often negatively skewed; for example, more old people tend to wear spectacles, have false teeth, or sustain heart attacks than young people.

7 List the four measures of dispersion

The range, the quartile deviation, the mean deviation, and the standard deviation.

8 Define the range

This is the difference between the largest and the smallest values in the distribution.

9 How is the range determined if the distribution is grouped?

Whether the first class is the same size as other classes or larger than the rest we take the midpoint of that group as the start of the range. If the last group is open-ended we assume it to be twice as large as the other groups and take the midpoint of that (i.e. one extra full class interval) as the end of the range.

10 What would be the range of salaries in Table 13.1 below.

Table 13.1 Salaries of Employees

Salaries (£)	Number of employees	Cumulative frequency
Under 3000	15	15
3000– 5999.90	37	52
6000– 8999.90	68	120
9000–11999.90	42	162
12000–14999.90	35	197
15000–17999.90	18	215
18000 and over	5	220
	220	

Answer: £1 500–£21 000.

11 What is the quartile deviation?

It is a measure of the spread of the data around the central item – the median. The quartiles are values which divide the data into four equal parts, the lower quartile at the 25 per cent level, the median at halfway, and the upper quartile at the 75 per cent level. The difference between the upper and lower quartiles is called the inter-quartile range, and the quartile deviation is half of the range (the semi-inter-quartile range). It therefore tells us to what extent the central half of the data is spread around the median position.

12 How do we find the quartiles of ungrouped data?

As with the median $\left(\dfrac{n+1}{2}\right)$ we use $\dfrac{n+1}{4}$ and $\dfrac{3(n+1)}{4}$. If the results come out to a fraction it is usual to express the quartiles to the nearest whole number.

13 What are the quartiles of the data set below?

Data set: 3, 5, 7, 13, 15, 16, 17, 19, 31. There are 9 numbers, so the median item is the value of the fifth item – which is 15. The lower quartile is $\dfrac{n+1}{4} = \dfrac{10}{4} = 2\frac{1}{2}$th item which is 2 to the nearest whole number. The second item is 5. The upper quartile $\dfrac{3(n+1)}{4} = \dfrac{30}{4} = 7\frac{1}{2}$th item. This is the 8th item, to the nearest whole number. The eighth item is 19.

14 So what is the interquartile range?

$19 - 5 = 14$.

15 So what is the quartile deviation?

The quartile deviation is half the inter-quartile range $= \dfrac{14}{2} = 7$.

16 So now describe the data set	It is a data set where the median is 15 and the spread of data around the median is ±7
17 How do we find the quartiles of a grouped distribution?	As with the median $\left(\dfrac{n}{2}\right)$ we use $\dfrac{n}{4}$ and $\dfrac{3n}{4}$.
18 Find the median and quartile items in Table 13.1 (page 48)	There are 220 items. Therefore: Median $= \dfrac{220}{2} = 110$th item

Lower quartile $= \dfrac{n}{4} = 55$th item

Upper quartile $= \dfrac{3n}{4} = \dfrac{660}{4} = 165$th item.

The median is in the £6000 plus group and is the 58th item in it. This is:

$$£6000 + \left(\frac{58}{68} \times £3000\right) = \underline{£8559}$$

The lower quartile is in the same group and is the third item in the group. This is:

$$£6000 + \frac{3}{68} \times £3000 = \underline{£6132}$$

The upper quartile is in the £12 000 plus group and is the 3rd item in the group. This is

$$£12\,000 + \frac{3}{35} \times £3000 = \underline{£12\,257}$$

19 What then is the quartile deviation of the set of data?

The interquartile range is:

$$£12\,257 - £6132 = £6125.$$

The quartile deviation is half this = £3062.50
To the nearest whole number this is
£3062

20 Now describe the data set

The salaries average £8559 snd the quartile deviation is ±£3062.

Written Exercise: *(a) What is the range of a set of data which starts with a first group of 'less than £100' wages and moves in £100 stages to '£900 and over'? (b) How would you describe a set of data that had a median value of 9850 tonnes, a lower quartile of 2685 tonnes and an upper quartile of 15 240 tonnes?*

Go over the topic again until you are sure of all the answers. Then tick it off on the check list at the back of the book.

14 Measures of dispersion – 2: the mean deviation

1 What is the mean deviation?

It is a measure of the dispersion of the individuals in a sample around the average chosen – which must be the arithmetic mean or the median.

2 How is it found?

We find the deviation of each item from the average of the whole set. Since some items will lie below the average and some above the average, we shall have negative and positive items. To avoid the deviations summing to zero we ignore the signs, treating them all as positive.

We can justify this by the argument that we only wish to find out how they are spread around the central position, not whether they are above or below the average. We now find the average deviation by dividing the aggregate deviations by the number of items.

3 This has been done in Table 14.1 below. Explain it

In Laburnum Close the sum of the deviations when we ignore the signs is 12 years, giving a mean deviation of 1.2 years. In Lilac Close the figures are 188 years and the mean deviation is 18.8 years.

Table 14.1 Ages of heads of households

Laburnum Close		Lilac Close	
Ages	Deviation from mean (40)	Ages	Deviation from mean (40)
36	−4	21	−19
39	−1	23	−17
39	−1	24	−16
40	0	24	−16
40	0	26	−14
40	0	28	−12
41	1	45	+5
41	1	62	+22
42	2	73	+33
42	2	74	+34
Sum (ignore signs)	12	Sum (ignore signs)	188
Mean deviation	1.2 years	Mean deviation	18.8 years

4 So compare the two sets of data using the measures of central tendency discovered

In Laburnum Close the heads of households are of very similar ages, with a mean age of 40 years and a dispersion around the average of only 1.2 years on average. In Lilac Close there is a much wider spread, with a mean age of 40 years and an average dispersion around that figure of 18.8 years.

5 Give the formula for the mean deviation of ungrouped data

To find the mean we use:

$$\bar{x} = \frac{\Sigma x}{n}$$

To find the mean deviation we use

$$MD = \frac{\Sigma(x - \bar{x})}{n}$$

but we ignore the signs of the deviations $(x - \bar{x})$

6 How do we deal with grouped data to find the mean deviation?

For grouped data the midpoint of each group is used to calculate the deviation from the mean, and the sum of the absolute deviations (i.e. ignoring the signs) is divided by the sum of the frequencies in the groups.

7 Give the formulae for these calculations

The formula for the arithmetic mean is

$$\bar{x} = \frac{\Sigma fx}{\Sigma f}$$

where x is the midpoint of the groups and f is the frequency in each group.
The formula for the mean deviation is

$$MD = \frac{\Sigma f(x - \bar{x})}{\Sigma f}$$

where x is the midpoint of the groups and $\bar{x}$ is the arithmetic mean found as shown above.

8 Why is the mean deviation not frequently used today?

Because disregarding of the signs is a fatal flaw for higher level statistical manipulation and a much better measure of dispersion, the standard deviation is available.

Written Exercise: *(a) What is a measure of central tendency? (b) What is a measure of dispersion? (c) How do these measures enable us to describe a set of data?*

Go over the topic again until you are sure of all the answers. Then tick it off on the check list at the back of the book

15 Measures of dispersion – 3: the standard deviation

1 What is the standard deviation?

It is a measure of dispersion of the samples around the centre point of a distribution. This centre point must be the arithmetic mean.

2 Why is it better than the mean deviation?

It is a better mathematical description since it does not ignore the sign of the deviation. Instead the deviations are squared (which eliminates the signs – they all become positive). After adding the squared deviations we find the average of the result. This is called the variance. We now take the square root of the variance to find the standard deviation.

3 Put this in mathematical terms – for ungrouped data

The sign for standard deviation is σ (little sigma). The formula is:

$$\sigma = \sqrt{\frac{\Sigma (x - \bar{x})^2}{n}}$$

$(x - \bar{x})$ are the deviations of each item from the mean. These are then squared to eliminate the minus signs, and added (Σ (big sigma) = the sum of). We then find the average by dividing by n, and then take the square root of the answer.

4 This has been done for one of the groups previously used (see Table 14.1 in Topic 14.) It is shown opposite in Table 15.1. Comment on the calculations

The mean age is 40 and we find the deviations from that mean. Some have negative signs and some positive signs. We square these deviations to eliminate the negatives. We add the squares of the deviations and find the average = 425.6. We now take the square root of the answer to get ±20.63 as the standard deviation.

Table 15.1 Ages of heads of household

Lilac Close

Ages	Deviations from mean (40)	Square of deviations
21	−19	361
23	−17	289
24	−16	256
24	−16	256
26	−14	196
28	−12	144
45	5	25
62	22	484
73	33	1089
74	34	1156
400		4256

Average of squared deviations (variance)	$= 425.6$
Square root of variance, i.e. standard deviation σ	$= \pm20.63$

5 The answer has come out slightly larger than the mean deviation (18.8 years). Why?

Because the process of squaring tends to emphasise the extreme items.

6 How is the standard deviation determined for grouped data?

We use an assumed mean and adjust our calculations for any errors caused by the difference between the assumed mean and the true mean.

7 Table 15.2 shows some data about hospital costs. Explain how we find the mean and the standard deviation.

(a) In column (ii) we have the midpoints of the classes. The only unusual one is the last one. Here we have assumed the open-ended class to be as big as three times the other classes, because some accidents are very expensive. (b) The total number of patients is 1504, but there are a lot of them in the lower cost

Table 15.2 Costs per hospital patient – casualty department

(i) Classes (£)	(ii) Midpoints of class	(iii) Number of patients (f)	(iv) Deviation (d¹) from mean	(v) Products f × d¹	(vi) Products f × d²
0– 49.9	25	87	−4	−348	+1392
50– 99.9	75	127	−3	−381	+1143
100–149.9	125	338	−2	−676	+1352
150–199.9	175	286	−1	−286	+286
200–249.9	225	134	0	0	0
250–299.9	275	98	1	+98	+98
300–349.9	325	62	2	+124	+248
350–399.9	375	41	3	+123	+369
400–449.9	425	71	4	+284	+1136
450–499.9	475	165	5	+825	+4125
500–549.9	525	42	6	+252	+1512
550–599.9	575	15	7	+105	735
600 and over	675	38	9	+342	+3078
Total		1504		+462	15474

brackets (838 of them had accidents costing less than £200). The average cost is expected to be a bit above this; so we assumed the mean to be at £225 – the midpoint of the next group. (c) We now take the deviations of the groups from the assumed mean, in class intervals. Thus group 4 is 1 class interval below £225 and is therefore −1. The final group is 9 class intervals above £225 (i.e. £450 above at £675). (d) We now multiply the frequencies in the groups by the deviation to get the product fd'. If we have assumed the correct mean, this column should sum to zero. It doesn't – it comes to +462 – so we have not assumed the correct mean. The correct mean is found by the formula

True mean = assumed mean
+ adjustment

$$= £225 + \frac{462}{1504} \times £50$$

where 50 is the class interval.
Therefore the true mean is
£225 + £15.36 = £240.36. (e) We now
square the deviations by multiplying
column (v) by column (iv) again. The
result is a set of positive deviations,
which sum to 15 474 class intervals.
Before we can find the standard
deviation by taking the square root of
the average of the squared deviations,
we have to adjust the figure of 15 474
by the square of the error made on the
assumed mean. So:

$$\sigma^2 = \frac{15\,474}{1504} + \left(\frac{462}{1504}\right)^2$$

$$= 10.288\,6 + 0.094\,4$$

$$= 10.383\,0 \text{ class intervals}$$

$$\therefore \sigma = \sqrt{10.383\,0} \times £50$$

$$= 3.222 \times £50$$

$$= \underline{£161.10}$$

8 So now describe the set of accident data clearly

The average accident costs £240.36
with a standard deviation around this
average of £161.10.

9 There is one disadvantage of all measures of dispersion. What is it?

They are all expressed in the units of the
original data. For example, in (8) above
the standard deviation is expressed in
sterling.

10 Why is this a difficulty?

We can only compare different sets of
data if they are expressed in the same
terms. For example we couldn't
compare data in sterling and data in
French francs.

11 How can we overcome this difficulty?	By using a **coefficient of variation**. A coefficient is just a number – not in absolute terms but in relative terms. To do this, we express the standard deviation as a percentage of the arithmetic mean.
12 What is the formula?	$$CV = \frac{100\,\hat{\sigma}}{\bar{x}}$$
13 Use this with the data of (8) above?	$$CV = \frac{100 \times 161.10}{240.36}$$ $$= \underline{67\% \text{ or } 0.67 \text{ of the mean}}$$
14 So now describe the set of data	The mean cost of an accident was £240.36 and the dispersion around the mean was 67 per cent of that figure.

Written Exercise: *The gross rateable value of dwellings (the value on which local taxes are levied in the United Kingdom) for 100 apartments rented unfurnished are as follows:*

Rateable value of dwellings (£)	Percentage of dwellings
Under 100	11
100 and under 200	32
200 and under 300	36
300 and under 400	8
400 and under 500	8
500 and over	5

a *Calculate the arithmetic mean and the standard deviation.*
b *Using your results describe the set of data. What desirable part of any description of a set of data are you unable to supply from the results?*

Go over the topic again until you are sure of all the answers. Then tick it off on the check list at the back of the book.

16 A comparison of the measures used to describe sets of data

1 Which is the best average?	The arithmetic mean.
2 Why?	Because it includes all the data in its calculation, and hence is influenced by every item.
3 When is it not the most useful?	(a) Where it is adversely influenced by an extreme item. (b) Where it is unrealistic – e.g. the average dog has 3.95 legs.
4 Which is the more appropriate measure in these circumstances?	The median – it at least is always central in the range of data.
5 Why is the mode of less use than the others?	(a) Because the most numerous items are not necessarily significant – for example in wealth statistics one millionaire outweighs many poor people. (b) Because you can have no-modal series, or bi-modal or tri-modal series.
6 Which is the best measure of dispersion?	The standard deviation, because it is mathematically sound, and can therefore be used in advanced statistical work.
7 Why is the quartile deviation frequently used?	Because it gives a reasonably good description of data which is very easy to calculate – the semi-inter-quartile range.

This is the end of Book 1 in *Revise and Test Statistics*. Another booklet, *Statistics 2*, is available for you to carry your revision work on further. It includes the following topics:

Measures of skewness
Time series analysis
Published statistics
Index numbers
Correlation and regression analysis
Rank correlation
Probability
Sampling and significance testing
Quality control

LONDON 1991

AM	LAW 1	JUNE 7
PM	LAW 2	June 11
AM	ECONOMICS	JUNE 17
PM	ECONOMICS	JUNE 20